The Natural History of The Ten Commandments

Ernest Thompson Seton

British Library Cataloguing-in-Publication Data
A catalogue record for this book is available from the
British Library

Ernest Thompson Seton

Ernest Thompson Seton was born on 14th August 1860, in South Shields, County Durham, England. He grew up to be a pioneering author, wildlife artist, founder of the Woodcraft Indians, and one of the originators of the Boy Scouts of America (BSA).

The Seton family emigrated to Canada when Ernest was just six years old, and most of his childhood was consequently spent in Toronto. As a youth, he retreated to the woods to draw and study animals as a way of avoiding his abusive father – a practice which shaped the rest of his adult life. On his twenty-first birthday, Seton's father presented him with a bill for all the expenses connected with his childhood and youth, including the fee charged by the doctor who delivered him. He paid the bill, but never spoke to his father again.

Originally known as Ernest Evan Thompson, Ernest changed his name to Ernest Thompson Seton, believing that Seton had been an important name in his paternal line. He became successful as a writer, artist and naturalist, and moved to New York City to further his career. Seton later lived at 'Wyndygoul', an estate that he built in Cos Cob, a section of Greenwich, Connecticut. After experiencing vandalism by some local youths, Seton invited the young miscreants to his estate for a weekend, where he told them what he claimed were stories of the American Indians and of nature.

After this experience, he formed the Woodcraft Indians (an American youth programme) in 1902 and invited the local youth to join (at first just boys, but later girls as well). The stories that Seton told became a series of articles written for the *Ladies Home Journal*, and were eventually collected in *The Birch Bark Roll of the Woodcraft Indians* in 1906. Seton also met Scouting's founder, Lord Baden-Powell, in 1906. Baden-Powell had read Seton's book of stories, and was greatly intrigued by it. After the pair had met and shared ideas, Baden-Powell went on to found the Scouting movement worldwide, and Seton became vital in the foundation of the Boy Scouts of America (BSA) and was its first Chief Scout (from 1910 – 1915). Despite this large achievement, Seton quickly became embroiled in disputes with the BSA's other founders, Daniel Carter Beard and James E. West.

In addition to disputes about the content of Seton's contributions to the Boy Scout Handbook, conflicts also arose about the suffrage activities of his wife, Grace, and his British citizenship (it being *an American* organization). In his personal life, Seton was married twice. The first time was to Grace Gallatin in 1896, with whom he had one daughter, Ann (who later changed her name to Anya), and secondly to Julia M. Buttree, with whom he adopted an infant daughter, Beulah (who also changed her first name, to Dee). Alongside his work with the Woodcraft Indians and the BSA, Seton also found time to pursue his primary interest – that of nature writing.

Seton was an early pioneer of animal fiction writing, his most popular work being *Wild Animals I Have Known* (1898), which contains the story of his killing of the wolf Lobo. He later became involved in a literary debate known as the nature fakers controversy, after John Burroughs published an article in 1903 in the *Atlantic Monthly* attacking writers of sentimental animal stories. The controversy lasted for four years and included important

American environmental and political figures of the day, including President Theodore Roosevelt. Seton was also associated with the Santa Fe arts and literary community during the mid-1930s and early 1940s, which comprised a group of artists and authors including author and artist Alfred Morang, sculptor and potter Clem Hull, painter Georgia O'Keeffe, painter Randall Davey, painter Raymond Jonson, leader of the Transcendental Painters Group, and artist Eliseo Rodriguez.

In 1931, Seton became a United States citizen. He died on 23rd October, 1946 (aged eighty-six) in Seton Village in northern New Mexico. Seton was cremated in Albuquerque. In 1960, in honour of his 100th birthday and the 350th anniversary of Santa Fe, his daughter Dee and his grandson, Seton Cottier (son of Anya), in a fitting tribute to the man who loved his surrounding countryside so much, scattered his ashes over Seton Village from an airplane.

The
Natural History
of the
Ten Commandments

By

Ernest Thompson Seton

Dedicated to
The Beasts of the Field
By a Hunter

The
Natural History
of the
Ten Commandments

The Ten Commandments

I. Thou shalt have no other gods before me.

II. Thou shalt not make unto thee any graven image, or any likeness of any thing that is in heaven above, or that is in the earth beneath, or that is in the water under the earth: Thou shalt not bow down thyself to them, nor serve them: for I the LORD thy God am a jealous God, visiting the iniquity of the fathers upon the children unto the third and fourth generation of them that hate me; And shewing mercy unto thousands of them that love me, and keep my commandments.

III. Thou shalt not take the name of the LORD thy God in vain; for the LORD will not hold him guiltless that taketh his name in vain.

IV. Remember the sabbath day to keep it holy. Six days shalt thou labour, and do all thy work: But the seventh day is the sabbath of the LORD thy God: in it thou shalt not do any work, thou, nor thy son, nor thy daughter, thy manservant, nor thy maidservant, nor thy cattle, nor thy stranger that is within thy gates: For in six days the LORD made heaven and earth, the sea, and all that in them is, and rested the seventh day: wherefore the LORD blessed the sabbath day, and hallowed it.

V. Honour thy father and thy mother: that thy days may be long upon the land which the LORD thy God giveth thee.

VI. Thou shalt not kill.

VII. Thou shalt not commit adultery.

VIII. Thou shalt not steal.

IX. Thou shalt not bear false witness against thy neighbour.

X. Thou shalt not covet thy neighbour's house, thou shalt not covet thy neighbour's wife, nor his manservant, nor his maidservant, nor his ox, nor his ass, nor any thing that is thy neighbour's.

MORE than one heathen philosopher conceived creation as a tree with its roots in the nether world, its fruit in the skies. Had these men been other than heathen, we to-day might have called them inspired. They outlined in advance the view of modern science, that the universe is an organic whole, a thing of growth, with ceaseless upward struggle.

Darwin and his school taught us the literal verity of this in material things.

Modern psychologists are daily discovering its truth in their own fields.

Possibly we may go further and find it apply equally in the moral world.

A theory is a great aid to study.

It helps one to observe, provided always one does not cut the facts to fit the theory, but rather keeps changing the theory to fit the new facts.

Years ago I set for my theory that: The Ten Commandments are not *arbitrary laws given to man,* but are *fundamental laws of all highly developed animals.*

If this be true I shall be able to trace them through the animal world. We can learn an unwritten law only by breaking it and suffering the penalty. My task

4

therefore was to discover among the animals disaster following breach of the ten great principles on which human society is founded.

There are two disasters commonly discernible: the first is, direct punishment of the individual by those he wronged; the other, a slow and general visitation on the whole race of the criminal, as the working out of the law. The former, the objective, is more obvious; the latter, the subjective, more important. But they are fundamentally the same, since the agents in the first case were impelled by their own recognition that wrong had been done, that a law had been broken.

Most commentators divide the Commandments into two groups:

The first four on man's duty to a Supreme Being.

The last six on man's duty to man.

For many reasons I found it better to take the latter group first, beginning with No. V.

V. Against Disobedience.

The law which imposes unreasoning acceptance of the benefits derivable from the experience of those over us. This is the foundation of all government, since the family is the social unit. Its force everywhere is so seen that it scarcely needs proof.

A Hen sets out with her Chickens a-foraging; one loiters, does not hasten up at her "cluck cluck" of invitation and command; consequently he gets lost and dies.

Another neglects to run to the spot when she calls in the established way that she has found "good food." He is not so well nourished as the others; he becomes a weakling, and in the first hard pinch he is the one that fails —he dies.

Again, she may call out "Hawk!" and run for shelter; the obedient ones run with her, and are safe; the disobedient loiter—and die. They pay the penalty, their days are short in the land.

Yet again: A Black-bear in the Cincinnati Zoo produced a family of two cubs in January, 1879. When they were seventy-one days old, one of them left the den for the first time, and followed the mother in her quest for food. This in a wild state would have been a fatal mistake for the young one. "As soon as the mother found it out," says Superintendent F. J. Thompson, "she immediately drove it gently back, and on the second attempt she cuffed it soundly, which put a stop to its wandering propensity. After a few days she allowed the cubs to wander about at will, provided no one was immediately in front of the den ; but so soon as a visitor put in an appearance, they were driven back into the den, and not allowed to emerge until the strangers were out of sight."

Under natural conditions this maternal rule was essential, and a breach of it meant death to the culprit.

When a mother Deer or Antelope sights, scents, or hears danger, she quickly communicates her warning to her young.

How it is done, varies greatly with the species; some bleat or snort; others may merely spread the disk of white hair around the tail, but all give what is understood to be warning of danger. The young at once squat in the grass, and the mother goes forth to baffle the foe as best she may. But it is essential to the little one and to the race that the warning be acted on promptly and fully.

This action on the part of the young is purely instinctive—which means that the law of obedience has been a long, long time in successful operation.

It would be easy to fill a volume with incidents illustrating this rule. But it is well known among all naturalists that obedience to parents is vital, and disobedience on the part of the young means injury to themselves, and, if uncurbed, death to the race.

VI. Against Murder.

That is, against taking the life of one of our own species. There is a deep-rooted feeling against murder in most animals. Their senses tell them that this individual is one of their own race, and their instinct tells them that therefore it is not lawful prey.

New-born Rattlesnakes will strike instantly at a stranger of any other species, but never at one of their own. I have seen a young Mink, still blind, suck at a mother Cat till fed, then try to take her life. Though a creature of such blood-thirst, it would never have attacked its own mother.

Wild animals often fight for the mastery, usually over a question of mates, but in practically all cases the fight is over when one yields. The vanquished can save himself either by submission or by flight. What is commoner than to see the weaker of two Dogs disarm his conqueror by grovelling on the ground?

The victor in a fight between two Cats is satisfied when the foe flies; he will not pursue him twenty yards. In either case had the enemy been of a different race the victor would have followed and killed him.

What makes the difference? Obviously not a reasoned - out conclusion, but a deep instinctive

feeling—the recognition of the unwritten law against unnecessarily
killing one's own kind.

There are doubtless exceptions
to this. Cannibalism is recorded
of many species; but investigation shows that it is rare except
in the lowest forms, and among
creatures demoralized by domestication or captivity. The higher
the animals are, the more repug-
nant does cannibalism become. It
is seldom indulged in except under dire stress of famine. Nothing but actual starvation induced
Nansen's Dogs to eat the flesh of
their comrades, although it was of-
fered to them in a disguised form.
Numberless experiences showed

me that it is useless to bait a Wolf-trap with a part of a dead Wolf. His kinsmen shun it in disgust, unless absolutely famished.

Obviously, no race can live by cannibalism; and this is instinctively recognized by all the higher animals. In other words, the law against murder has been hammered into animals by natural selection, and so fully established that they will not only abstain from preying on one of their own tribe, but will rally to rescue one whose life is threatened.

The fact that there are exceptional cases does not disprove the law among beasts any more than among men.

15

VII. Against Impurity.

Although on the face of it directed against the grossest form of misapplied reproductive instinct, most commentators agree that it is meant to cherish the general principle of purity.

Of what service is such a general principle to the race? A review of many creatures and their marriage customs shows that from the beginning they have been groping for an ideal form of marriage.

Promiscuity was doubtless the mode when first sex appeared in the animal world. It had the great advantage that it insures all finding mates with whom fruitful union is possible. But it has several disadvantages, the most obvious being that unlimited personal contact opens the way for epidemic diseases of all sorts. The less personal contact, the less disease.

The promiscuous animals to-day —the Northwestern Rabbit and the Voles—are high in the scale of fecundity, low in the scale of general development, and are periodically scourged by epidemic plagues.

The Chinaman who reduces personal contact to a minimum by refraining even from shaking the hand of a friend, has gone to the extreme, and without doubt has had his reward.

Another danger from this lawless reproduction is the evil called "inbreeding," that is, the mating of near kin.

Promiscuity has been displaced by polyandry and polygamy, among certain animals. That the former has not been a success is shown by the fact that it is very rare among the higher kinds, and practised only under exceptional circumstances.

The few cases I can find are the European Cuckoo, and, possibly, the American Cowbirds. The extraordinary, hazardous and dishonest methods these are driven to for support of their young are well known.

The fact that these species are

healthy and prospering is a puzzle to me. Nevertheless it must be observed that their parasitism is *on the other races, not on their own kind.*

Polygamy seems much more satisfactory: there are hundreds of species of polygamous animals in the world to-day that are prospering and growing with the world's growth.

On the face of it, polygamy might seem to be good, because it makes it possible for only the finest males to breed, and insures for them the greatest possible number of offspring.

This sounds convincing, but some unexpected light has been shed by

Caton's observation among the Wapiti, the most polygamous of all our
Deer.

Referring to Sultan, the great bull
Wapiti that for a longer time than
any other was the monarch of the
herd in his park, he says:*

"At first his progeny were rea-
sonably numerous, but during the
last few years of his life they grad-
ually diminished from a dozen to
a single fawn in 1875, with about
twenty-five females, more than half
of which had previously produced
fawns." He was removed, though
yet able to hold the harem by force,
and replaced by a younger buck;
"the result was that I had twelve

* Antelope and Deer of America, pp. 294-5.

fawns the next season, including one pair of twins." It is probable that a far better result would have been secured had each female been paired off with a single male.

As the Wapiti is the most polygamous of the Deer in America, probably in the world, it is interesting to note that it is the first of the family to disappear before civilization. This may be due in part to its size; but it is further remarkable that the most successful of all our true Deer, that is, the common White-tail, is the least polygamous.

There is at least one strong and obvious objection to polygamy among animals: the offspring of such union have but one parent to care for them, and the weaker one at that.

It is commonly remarked that while the Mosaic law did not expressly forbid polygamy, it surrounded marriage with so many restrictions that by living up to the spirit of them the Hebrew was ultimately forced into pure monogamy.

It is extremely interesting to note that the animals in their blind groping for an ideal form of union have gone through the same stages and have arrived at exactly the same conclusion. Monogamy is their best solution of the marriage question, and is the rule among all the highest and most successful animals.

There are four degrees of monogamy:

One, in which the male stays with one female as long as she interests him or desires a mate, then changes to another; for his season may be many times as long as hers. Thus he may have several wives in the season, but only one at a time. T' is convenient for both parties, .i it is open to the same objection as frank polygamy. It is the way of the Moose.

A second kind, in which the male and one female are paired for that breeding season only, the male staying with the family, and sharing the care of the young till they are well grown; after which the parents may or may not resume their fellowship. This is admirable. It is seen in Hawks.

A third, in which the pair consort for life, but the death of one leaves the other free to mate again. This is ideal. It is the way of Wolves.

A fourth, in which they pair for life, and in case of death the survivor remains disconsolate and alone to the end. This seems absurd. It is the way of the wild Geese.

Upon the whole we find the animals succeeding, that is, avoiding disease and holding their own, spreading, and high in the scale, in proportion as they approach the ideal union.*

I confess, however, that monogamy in the fourth degree puzzles me.

In making observations, one is hampered by the fact that association with man has always been ruinous to the morals of animals.

There can be no doubt that the Dog, now so promiscuous, was originally a monogamous creature. One of the great difficulties besetting the growing of Blue-foxes for their fur, on the islands of the Behring Sea, is what has been

---

* Dr. Woods Hutchinson in "Animal Marriage" has pointed out that other things being equal, a monogamous race will beat a polygamous c .e in the long run.

25

called the obstinate and deplorable monogamy of those animals. The breeders are working hard to break down this high moral sentiment and produce a Blue-fox that does not object to polygamy, promiscuity, or any other combination, and so remove all sentimental obstacles to their experiments.

The wild Goose is a most exemplary bird; the tame Goose is little better than the Dog. Of Rabbits, wild or tame, the less said the better.

There is, however, one domestic bird that maintains its honorable wild tradition in spite of all that sinful man can do ; that is the Pigeon. The breeder knows that

the young in a given nest are un-
questionably the offspring of their
alleged parents, no matter how
many hundreds of their kind may
freely fly with them all day.

What wonder that Gadow, the
distinguished ornithologist, should
proclaim the Pigeons the birds of
the future, implying that when,
under the relentless unwritten laws,
all other species shall have paid the
penalty and run themselves out, the
Pigeons will be happily possessing
the earth.

Similarly the most successful wild
quadrupeds in American to-day are
the Gray-wolves. Not only have
they through strict monogamy
eliminated much possibility of dis-

ease, and given their young the advantage of two wise protectors, but they have even developed a spirit of chivalry; that is, the male shows consideration for the female in the non-mating season on account of her sex. This is very high in the scale. And one result, at least partly due to these things, is that the Wolves defy all attempts to exterminate them, and are increasing to-day in exact ratio to the improved food supplies for which the settlers are responsible.

The proverbial exceptions undoubtedly occur, and they have their value as proof, not disproof.

Immorality in its broadest sense may be defined as the deflection of any natural power, member, or instinct from its proper purpose to one that works harm for the species.

Among animals we have recorded nearly every kind of abominable vice that was known among men, and forbidden by the Mosaic law.

In captivity and domestication we see such things all too often, but rarely in a state of nature, partly because the cases are scarce and difficult to observe, and partly because the creatures of vice soon destroy themselves; they pay the extreme penalty.

Incest is admittedly forbidden by the spirit of this ordinance. The numberless contrivances among plants to prevent any but cross-fertilization, evidence the importance of preventing the marriage of near kin. Among higher animals, strange to tell, observation of this law is not so marked, probably because their safeguard is not a mechanism, but a sentiment, which suffers in domestication and in captivity. It seems to exist, however.

Mr. L. H. Ohnimus, for years the director of Woodward's Garden Menagerie at San Francisco, told me that often among higher animals they had great difficulty in mating brother and sister that were brought up together. The friendly feeling commonly overpowered the sex instinct. If, however, the pair were separated long enough to be brought together as practically strangers of opposite sexes, the difficulty disappeared.

But the penalty must be paid. The resultant young in most cases are feeble creatures, tending to die out in a generation or two, that is, paying with their death for the sin of their parents. This is physical law, and the fact that it was unwitting sin does not in any degree absolve the sinners from the consequences.

To sum up: There is evidence that in the animal world there has long been a groping after an ideal form of marriage. Beginning with promiscuity, they have worked through many stages into pure monogamy; and, other things being equal, the species, owing to natural laws, are successful in proportion as they have reached it, and therefore have developed an instinctive recognition of the seventh commandment.

VIII. Against Stealing.

The whole property question is in this, and the high development of the property idea among animals must be a surprise to all who have not studied it. This is the animal law:

The producer owns the product; unproduced property belongs to the first who discovers and possesses it.

Numberless instances in proof will occur to every naturalist. Property among animals consists of food, nest, playground, range, and wives. Ownership is indicated in two ways: one by actual possession, the other by ownership marks. Of these there

33

are two kinds, smell marks and visible marks; by far the more important are those of smell.

I once threw peanuts for an hour to the Fox-squirrels in City Hall Park, Madison, Wisconsin. In each case, the peanut, when thrown, was no one's property. All the near Squirrels rushed for it; the first one to get it securely in his mouth was admittedly the owner; his claim was never questioned after a few seconds' actual possession. If hungry he ate it at once; otherwise his first act was to turn it round in his mouth three or four times, as he licked it, marking it with his own smell, before burying it for future use.

This is paralleled in many tribes of men. Eskimo of Davis Strait, according to Franklin, lick each new acquisition by way of taking possession. Sailors commonly spit on a new-got article, and boys, in the north of England at least, indicate the beginning of their ownership in the same way. Many animals, as Rabbits and Bears, rub their bodies against trees in their range, to let other animals know that this place is already possessed. Some creatures, as the Weasels, have glands that secrete an odor which they use for an owner-mark. As this odor must vary with each individual it answers admirably. I have seen

Martens, Wolves, and Foxes marking their property in this way. The Wolverine is commonly described as a monster of iniquity, that not only lugs off and hides the hunter's food, but defiles it with his abominable secretion, so that it is useless to the original owner. It is quite certain that malice of this kind is unusual; although Dogs and Wolves, high in mental development, have been observed to show scorn in this manner. The Wolverine eats what he can of the trapper's hoard, and hides the rest for future use, after taking care to mark it with his ownership smell-mark.

Foxes and Wolves are known to

store up food, and after it is buried they defile the place in a characteristic way. Many harsh terms are applied to this practice. It is, or was formerly, ascribed to the inherent and abominable filthiness of all creation unregenerated in the particular manner specially advocated by the then critic. The fact is that the odor glands of the Fox and Wolf are so situated that their product is given out with the product of the kidneys. They do this, then, merely to put their mark on their cache.

Thus they have the property instinct in high development.

In the August of 1906, at Petoskey, Mich., I made the acquaintance of a team of Eskimo train Dogs — they were seven-eighths Wolf, and showed all the wild traits in force. The leader, a big savage creature, was easily master of the others. I gave the smallest one a bone after he was already fed. True to the wild instinct of his kind, he set off to hide this for future use. The bone was buried under · cedar bush some hundred yards away, and the place marked in Dog fashion. The owner then retired about fifty yards to a shady spot, where he could see his cache, and lay down.

The biggest Dog of all saw the hiding of the bone, but did not see the watcher. He walked quictly to the cache. When within twenty feet, there could no lonrer be any doubt of his purpose; the smaller Dog rushed from his covert and stood guard over his property, showing his teeth and clearly intimating that only over his dead body could the bully take his property. The big Dog, though he could have whipped the smaller in a minute, turned slowly and sullenly away, as though he knew his cause was weak.

What is the psychology of this situation? (And it was purely psychological.)

Can any one deny that the little Dog felt that he was right, the big Dog that he was wrong? In other words, they recognized the law of property, and that stealing was crime.

Many instances of this kind could be adduced. The principle is very old, and has, indeed, given rise to several proverbs: "Any cock will fight on his own dung-hill"; " He is a poor thing that won't fight for his own "; " Thrice is he armed that hath his quarrel just," etc.

For how long are these caches made? In the case of domesticated Wolves they are opened and the contents eaten within a few hours or days at most. But I found it the opinion of hunters, that among the truly wild animals the cache is made in time of plenty for a season of starvation, maybe months ahead.

There is good reason for believing, however, that the Wolf, Coyote and Fox have no compunction about stealing from each other. I found it a most alluring bait, if I buried a piece of meat, that is, formed a cache, and either made it fair game for Wolves by pattering the ground with an old Coyote foot, or leaving it with man tracks only around. Whether pattering it with a Wolf's foot would make other Wolves respect it, I am not prepared to say.

The food idea is probably the first property idea. Ownership of the home-place came later, but is now deeply rooted.

Many cases in line have been reported to me from among rookeries in England. Rooks are ordinarily moral birds. A stick found in the woods is the property of the Rook that discovers it, and doubly his when he has labored to bring it to his nest. This is recognized law. Nevertheless there are degenerates or thieves that think it easier to steal sticks from their neighbor's nest than to fetch them from afar. The result is war.

In the autumn I put up opposite my window an artificial shelter hole for birds. A Flying-squirrel used it for a nest. In the spring I several times saw a pair of Chickadees peeping into the hole, but noting the nesting material, the evidence of a possessor, they withdrew without entering. If they knew that the occupant was a Squirrel, fear may have kept them back, and the incident means nothing; but all they could see were some shreds of bark which might have represented the nest of another Chickadee, in which case they were restrained by the unwritten law.

To get without labor is theft; and the thief and his children must be the sufferers in the end. ·

How does this work out in our animal world?

The Squirrel that will not store

43

must starve or steal in winter. If he escapes being killed by his honest neighbors, the vice of stealing will spread, so that it will no longer be worth while to store up for winter, and the habit will be abandoned.

We must remember that the lives of animals are in a delicate balance; at times a featherweight easily turns the scales against them. A single hard winter among Squirrels that had been forced to abandon storage, might wipe out the whole race.

So also among Rooks. The thief taken red-handed may suffer grievous bodily punishment, or even death; this is the objective retribution. But the subjective is farther reaching, for a spread of the vice would prove ruinous to all the nests, and tend to exterminate the race.

Out of the food-property instinct has grown the territory-property instinct. Bears, Martens, Foxes, Wolves, and many other species mark their range by putting their signs on trees, stones, etc., scattered over the region claimed.

Bears not only rub their backs on the trees, but claw them and tear them with their teeth. These things are familiar to all who have lived among Bears. The visible marks may appeal to the eyes of another Bear when he is far off, but the smell record is, I take it, of chief

45

importance, and is the only one used by Wolves and Foxes.

These are the marks of ownership: to what extent are they respected?

It is well known that each wild animal has a little home region or range that he considers his, and for which he will fight. But it is not so well known that others of his kind will respect his claim without any fight, without anything, apparently, but the little sign-boards or smell-marks already noted. Dr. F. W. True, writing of the Blue-foxes on the islands of the Behring Sea and their tameness, says one of them will follow a man for a long way, apparently hoping to be fed, will follow indeed "to the boundary of his domain, for each Fox, like his neighbor, the bull Seal, seems to have a definite territory . . . which he regards as his own, and upon which he resents intrusion." *

From these examples it will be seen that the operation of natural laws has produced in the animals ideas of property rights in materials and in places, and means of putting those rights on record. That is, has tended to give ever-growing force to the law against stealing.

* Fox Propagation in Alaska, Rep. Sec. Int., 1903, p. 80.

IX. Against False Witness.

Although the commandment forbids especially false witness against a neighbor, it is generally considered to have a broader meaning—to prohibit any falsification.

In Fox-hunting the character of every Hound becomes well known, not only to the men, but to the Hounds themselves. When they are scattered for a "find," each Hound does his individual best and is keen to be first. Oftentimes a very young Hound will jump at a conclusion, think, or hope, he has the trail, then allowing his enthusiasm to carry him away, give the first tongue, shouting in Hound language, "Trail!" The other Hounds run to this, but if a careful examination shows that he was wrong, the announcer suffers in the opinion of the pack, and after a few such blunders, that individual is entirely discredited. Thenceforth he may bawl "Trail!" as often as he likes, no one heeds him.

The spread of such a habit of false witness would be disastrous to the whole race of Dogs in a wild state. They would discredit each other. All the enormous benefits derivable from collaboration would be lost to them; and since it takes but a little thing long continued in the struggle for life to work great changes, it is easily conceivable that this vice of lying might exterminate the race that became addicted to it.

The wild animals no doubt afford safer instances, but they are so difficult of observation that few are at hand. One of the most remarkable cases in point is among Wolves. I do not know that the incident is true, but it sounds true, and there is no inherent reason why it should not be so. The story appeared in the "Leisure Hour" in the volume of 1892–3, and was written by E. L. Hickey.

It was many and many a league away
    from the place where now we are,
And many a year ago it happ'ed in the
    land of the Great White Czar.

It was morn—I remember how cold it
  felt—out under a low pale sky,
When we moored our boat on the river
  bank, my companion Leigh and I.
And the plunge in the water unwarmed
  of the sun was less for desire than
  pluck,
And we hurried on our clothes again
  and longed for our breakfast luck ;
When all of a sudden he clutched my
  arm and pointed across, and there
We stood up side by side and watched,
  and as mute as the dead we were.

We saw the gray-wolf's fateful spring,
  and we saw the death of the deer.
And the gray-wolf left the body alone,
  and swift as the feet of fear
His feet sped over the brow of the hill,
  and we lost the sight of him
Who had left the dead deer there on the
  ground uneaten, body or limb.

So when he vanished out of our sight
    we rowed our boat across,
And lifted the carcass and rowed again
    to the other side.   The loss
For you, good Master Wolf, much more
    than the gain for us will be.
'T were half a pity to spoil your sport,
    except that we fain would see
The reason why with hunger unstanched
    you have left your quarry behind ;
Red-toothed, red-mawed, foregone your
    meal ; Sir Wolf, we'll know your
    mind.

Hungry and cold we waited and watched
    to see him return on his track ;
At last we spied him atop of the hill,
    the same gray-wolf come back,
No longer alone, but a leader of wolves,
    the head of a grewsome pack.
He went right up to the very place
    where the dead deer's body had lain,

And he sniffed and looked for the prey
  of his claws, the beast that himself
  had slain.
The deer at our feet and the river between and the searching all in vain.
He threw up his muzzle and slunk his
  tail and whined so pitifully,
And the whole pack howled and fell on
  him—we hardly could bear to see.
Breaker of civic law, or pact, or what-
  ever they deemed of him,
He knew his fate and he met his fate,
  for they tore him limb from limb.

I tell you we felt as we ne'er felt since
  ever our days began—
Less like men that had cozened a brute
  than men that had murdered a man.

This, of course, was a tragic mis-
carriage of justice, but the princi-

ple is well known. All the higher animals profit by each other's knowledge through methods of intercommunication. Falsification would certainly work dire disaster.

X. Against Coveting.

The broad principl_ _f th_s _mmandment _ ag_inst _du _ nkering f_r a _ _ _bo_ p_ _ _ against _chen __ to _is_ him.

A remark _b_ _ ha _ cu _ many times _ l_ _ n t_ cou_.try around Yellow_t_ _e _ It may have pre__t a _licati_ _.

A ban_ _ _pit_ _ _ing southward to th_ir wi__ _ng came on _e _aystack _ _ion__r. It _ _ _ce_ in _ _t _ ey could n _ _ _ _r. _t _ sm_lt so de-_ _a _ t_ _ _d lingered abou_ _ _n_ time to get possession. Thus the days passed, the Deer grew weaker, winter came down, and the whole band perished; whereas, had they moved on or worked to find their proper food they would, as often before, have come safely through to the spring.

In this case I am by no means sure of the principle involved, and cite the incident with much hesitancy. A weak spot in the illustration is the circumstance that the possessor of the stack was *not another Elk*.

A more nearly pertinent circumstance was recently told me by a friend.*

* Mr. H. Dallas, of Morristown, Ohio.

Under the barn eaves at his home a colony of Swallows had for long been established. In the spring of 1885 a pair of Bluebirds came and took forcible possession of one of the nests. The owners first tried to oust the invaders, next the whole Swallow colony joined in the attempt, without success. The Bluebird inside was entrenched behind hard mud walls, and defied them. At length the Swallows came in a body, each with a pellet of mud, and walled up the entrance to the nest. The Bluebird in possession starved to death, and was found there ten days later.

In this case the retribution came

direct from the Swallows, in obedience to the inner impulse. But it
is clear that Bluebirds adopting
habitually these methods of nesting would become parasites de-
pendent on the Swallows; this ad-
ditional burden might easily turn
the balance of nature against the
Swallows, ending in their death
as a species and, of course, the
death of their dependents.

A still more obvious episode I
have seen many times in the barn-
yard. A Hen had made a nest in
a certain place, and was already
sitting. Later another Hen, de-
siring the same nest, took possession several times during the owner's brief absence, adding some of

her own eggs, and endeavoring to sit. The result was a state of war, and the eggs of both Hens were destroyed.

It is not easy to say whether this was coveting or stealing, but I find it equally difficult to discriminate between the two laws that forbid these things.

This was the last of the lower group of commandments, and here my pathway seemed to end. If the next in order merely enforced a period of rest among toilers, then could I find illustrations among all toilers. But this would be a physical interpretation, and would take it out of the superior class of ordi-

nances, where commentators generally agree that it belongs. They
maintain that its purpose is to
set apart a time for spiritual
matters, and of this there was no
discernible recognition in my field.
I could find nothing in the animal world that seemed to sug-
gest any relation to a Supreme
Being.

Therefore I reformed my theory
to fit the new facts, and presented
it thus:

The first four commandments
have a purely spiritual bearing;
the last six are physical. Man is
concerned with all, the animals
only with the last six.

I was also struck by the thought

that in all cases the ultimate penalty is death.

There was another, a disappointing conclusion forced on me. It seems that law exists only between members of the same species. Wolf and Wolf have law, Crow and Crow, Weasel and Weasel, Mouse and Mouse even, but never so far as I can see, Wolf and Mouse, or Crow and Weasel. There is nothing but bitter war between them; their might is their right.

We should not marvel at this, however, since it was ever thus with man until the latest light came. Ask any savage which is worse, to steal some trifling article, the property of his tribesman, or to massacre a family of the neighboring tribe. He will as surely answer the former as we should the latter.

Only in his highest development is man capable of the broad love and sympathy that take in all the human race, and extend even to the beasts of the field.

With this conclusion then I was forced to halt the investigation: That we may find in the animals the beginnings of man's physical and mental attributes, but not a vestige of foundation for his spiritual nature. And the conclusion seemed the end. Because the trail became obscured I thought it went no further. But a faint glimmering of light came unexpectedly.

My twenty-five years of journals had been copied and the copies cut up so that incidents referring to each subject might easily be filed. I found several new subjects well represented, such as the evolution of sanitation, amusement, intercommunication, etc., and a final department of *unexplained strange instances*; when I got many of these together I found that they began to explain each other. To make this clear I give several of them now:

1st. Dr. G. B. Grinnell tells me that when out shooting with General Custer's party near the Black Hills in 1874, they observed a Falcon in pursuit of a wild Pigeon; when the latter saw that it could not escape its winged foe, it took refuge among the men, resting on one of the saddles.

2d. Mr. Geo. F. Guernsey, of Fort Qu'Appelle, Saskatchewan, writes me that some years ago a neighbor and his wife standing in their cattle yard saw a pack of five Coyotes chasing a Fox. The Fox was pretty nearly spent; it ran finally right up to the woman, and crouched for protection at her feet.

3d. In the December of 1886, I was hunting Snow-shoe Rabbits in a little grove near Carberry. The one I was pursuing escaped. It was an exceedingly cold day, some 35 degrees below zero. I laid my gun on my sleigh and busied myself lighting a fire to make some tea. As I cowered over this trying to think I was getting warm, I saw a Rabbit running through the little grove. It ran past me some forty yards away; then I noticed some twenty-five feet behind it another Rabbit running very fast in pursuit. The first circled round, came nearer. Now I saw that the smaller Rabbit was not a Rabbit at all, but a white Weasel, an Ermine, that was running the Rabbit down. The chase continued around me, but ever nearer. Though so much swifter the Rabbit was losing because the paralysis of terror was setting in.

The Weasel was within a few feet of his victim and ready for the final spring, when that Rabbit made a rush toward me, and took refuge under the sleigh near my feet— *came to me*, who had been trying to kill it a few minutes before.

The Weasel flashed about and under the snow, curling his nose a little; then realizing that he was probably running into danger, darted under brush and snow to vanish. The Rabbit cowered at my feet for a few minutes, but recovered and hopped away in another direction.

4th. In the October of 1898, I was riding across the Bighorn Basin (Wyoming) with Mrs. Seton and

Mr. A. A. Anderson, when we noticed near the horizon some bright white specks. They were moving about, disappearing and showing again. Then two of them seemed to dart erratically over the plain, keeping always just so far apart. Soon these left the others and careered about like twin meteors, this way and that, then our way; at first in changing line, but later directly toward us.

Their wonderful speed soon ate up the intervening mile or two, and we now saw clearly that they were Antelope, one in pursuit of the other. High over their heads a Golden Eagle was sailing.

On they came; the half-mile

shrank to a couple of hundred yards, and we saw that they were bucks, the hinder one larger, dashing straight toward us still. As they yet neared we could see the smaller one making desperate efforts to avoid the savage lunges of the big one's horns, and barely maintaining the scant six feet that were between him and his foe.

We reined up to watch, for now it was clear that the smaller buck had been defeated in battle with an exceptionally vicious rival, and was trying to save his life by flight. But his heaving flanks and gaping, dribbling mouth showed that he could not hold out much longer. Straight on he came toward us, the deadliest foes of his race, the ones he fears the most.

He was between two deaths—which should he choose? He seemed not to hesitate—the two hundred yards shrank to one hundred, the hundred to fifty—then the pursuer slacked his speed. It would be folly to come farther. The fugitive kept on until he dashed right in among our startled horses. The Eagle alighted on the rock two hundred yards away.

The victorious buck veered off, shaking his sharp black horns and circling at a safe distance around our cavalcade to intercept his victim when he should come out the other side. But the victim did not come out. He felt he was saved, and he stayed with us. The other buck seeing that he was balked, gave up the attempt, and turning back, sailed across the plain till he became again a white speck that joined the other specks, no doubt the does that had caused the duel.

The vanquished buck beside us stood panting, with his tongue out, and showing every sign of dire distress. It would have been easy to lasso him, but none of us had any desire to do him harm. In a very short time he regained his wind, and having seen his foe away to a safe distance, he left our company to go off in the opposite direction.

The Eagle realized now that he was mistaken in supposing that something was to be killed, and that there would be pickings for him. He rose in haste and soared to a safe distance.

5th. This I heard from George Crawford, the well-known guide of Mattawa; it was corroborated by others in camp:

In March, 1888, while out with his partner to catch Moose for Dr. S. Webb, he came on a Moose-calf track in the deep snow. There was no sign of a cow, so they turned their Dog loose. Very soon they heard him barking, and came up to the calf. It rushed toward them with bristling mane. His partner ran away, and he got behind a tree. The calf charged up to him and quickly wheeled to face the Dog. It paid no heed to the man then, but when he turned homeward it followed him for protection, crowding up close and watching the Dog. At home he put a halter on it, and it allowed him to lead it quietly into the stable. It was shipped to Dr. Webb, and is now roaming the Adirondacks.

6th. The following was related to me by Edouard Crête, of Deux Rivières:

In late September of 1893, a mail-carrier was starting from Bear Lake to Deux Rivières. Crête showed him a short cut over Brulé

Lake. Some hours later two men were out that way looking for axe-handles, and heard the mail-carrier shouting for help. Instead of going to him they ran back to camp in great fear. The foreman picked up a rifle and, accompanied by Crête, went as fast as possible to the place. They heard the shouting as soon as they came within a half-mile. When near enough, he called out: "A Moose has got me up a tree." They came close, and saw it was a cow Moose. She would neither go away nor charge. Indeed, she paid no attention to them. The foreman, Jean Basquin, walked up within twenty yards and shot her.

The mail-carrier, it seems, had come on the cow suddenly. She was alone, but came toward him squealing. Her mane was up, and she seemed to be threatening him. He had nothing but a hatchet, so ran for a tree, and happened to find one leaning so much that he could walk up. She ran behind him within touching distance all the way, but did not strike at him. The tree at the highest point was only ten feet up. Here the man sat, the Moose below. She could easily have struck him, but made no attempt to do so. There she stayed watching him; her mane bristled all the time.

When she heard the other men

74

coming she merely turned her head, but during the three hours that she kept the man up that tree she did not leave the spot for a moment.

When examined after skinning, her left side was found in a dreadful condition. Evidently she had been attacked by a bull Moose some days before. The horns had pierced her flank in five places. The side was all inflamed and matter had formed in four places. She must have been suffering great pain, and would surely have died before long. They could not make out why she should go to the man, but it is quite certain she was not there to do him any harm, for she had every opportunity and did not strike at him once.

Why then the angry bristling of her mane? Perhaps it was not anger. It may have been any other intense feeling. It is not easy to discriminate so finely the expressions of animal emotion. We only know that she was greatly wrought up about something.

These are the incidents. They seem to have a common principle. Divested of externals, what is the cardinal thought in each? This, I take it—that when the animals are in terrible trouble, when they have done all that they can do, and are face to face with despair and death,

there is then revealed in them an
instinct, deep-laid—and deeper laid
as the animal is higher — which
prompts them in their dire extremity to throw themselves on the
mercy of some other power, not
knowing, indeed, whether it be
friendly or not, but very sure that
it is superior.

Here perhaps is the looked-for
light. I was seeking in the animal
nature for beginnings of the spiritual life in man, for something that
might respond to the four higher
ordinances. Maybe in this instinct
of the brute in extremity, we have
revealed the foundation of something which ultimately had its high-

est development in man, reaching, indeed, like the Heathen Thinker's Tree, from root in the earthy darkness to its fruit in the Realm of Light.

www.ingramcontent.com/pod-product-compliance
Lightning Source LLC
Chambersburg PA
CBHW031241260626
47169CB00007B/2399